YOUR KNOWLEDGE HAS VALUE

A new perspective on the determination of the prime numbers

William Fidler

GRIN ☺

Bibliographic information published by the German National Library:

The German National Library lists this publication in the National Bibliography; detailed bibliographic data are available on the Internet at http://dnb.dnb.de.

ISBN: 9783346882721
This book is also available as an ebook.

© GRIN Publishing GmbH
Trappentreustraße 1
80339 München

Print and binding: Books on Demand GmbH, Norderstedt, Germany
Printed on acid-free paper from responsible sources.

The present work has been carefully prepared. Nevertheless, authors and publishers do not incur liability for the correctness of information, notes, links and advice as well as any printing errors.

GRIN web shop: https://www.grin.com/document/1360641

A new perspective on the determination of the prime numbers

W M Fidler

Abstract

A procedure is developed here for determining the primality of a number, N, but does not examine the number, but rather, a function of that number.

The function has several properties by means of which many of the non-primes may be identified and discarded without further investigation.

Using the concept of cage numbers (defined in the text) a matrix of all of the counting numbers is produced, where, with the exception of the prime numbers 2 and 3, all of the prime numbers are embedded in one column of the matrix. Strings of consecutive numbers are extracted from the matrix, and it is seen that the only places within the whole of the range of the counting numbers where prime numbers can exist are at the positions of the second and penultimate numbers within a string. By means of the function of N, all of the odd natural numbers may be generated and examined for primality.

2

List of Contents

Introduction

As is well known, beginning with Euclid, who showed that the number of prime numbers is infinite, the study of the prime numbers has held an abiding fascination over millennia.

The importance of identifying a prime number is demonstrated by the statement of Veisdal [1], when after quoting a string of prime numbers ending in 43 he makes the following statement, which we quote verbatim

'Present an argument or formula which (even barely) predicts what the next prime number will be (in any given sequence of numbers) and your name will be forever linked to one of the greatest achievements of the human mind, akin to Newton, Einstein and Gödel'.

Analysis

In order that the reader may have sufficient data from which he/she may test the methods developed here we present sets of data taken from [2].

If we stack the odd number line in the manner shown below it is seen that we have formed a matrix having 15 columns and, assuming that the odd numbers extend to infinity, will have an infinite number of rows. The numbers shown underlined and in bold are prime numbers obtained using a prime number calculator available on Wikipaedia. Inspection of the array shows that, with the exception of columns which have a prime number as their first entry, there can never be prime numbers in columns 1, 2, 4, 7, 10, 12, and 13. Any number in the array (and the extension thereof to infinity), the sum of whose digits is divisible by 3 is not prime and can be rearranged to yield other numbers which cannot be prime; in the context of this work, which is entirely devoted to the prime numbers, any rearranged number must not end in an even number and any number, rearranged or otherwise which ends in a 5 is a member of one of columns 2, 7, or 12 and is not prime. It should be noted that the 'spacing' between adjacent rows is 30, whilst that between adjacent columns is 2.

3	**5**	**7**	9	**11**	**13**	15	**17**	**19**	21	**23**	25	27	**29**	**31**
33	35	**37**	39	**41**	**43**	45	**47**	49	51	**53**	55	57	**59**	**61**
63	65	**67**	69	**71**	**73**	75	77	**79**	81	**83**	85	87	**89**	91
93	95	**97**	99	**101**	**103**	105	**107**	**109**	111	**113**	115	117	119	121
123	125	**127**	129	**131**	133	135	**137**	**139**	141	143	145	147	**149**	**151**
153	155	**157**	159	161	**163**	165	**167**	169	171	**173**	175	177	**179**	**181**
183	185	187	189	**191**	**193**	195	**197**	**199**	201	203	205	207	209	**211**
213	215	217	219	221	**223**	225	**227**	**229**	231	**233**	235	237	**239**	241
243	245	247	249	**251**	253	255	**257**	259	261	**263**	265	267	**269**	271
273	275	**277**	279	**281**	**283**	285	287	289	291	**293**	295	297	299	301
1	2	3	4	5	6	7	8	9	10	11	12	13	14	15

Further, in order to extend the data available to the reader we present another part of the matrix, which, solely for reasons of space does not appear in tabular form.

2013 2015 **2017** 2019 2021 2023 2025 **2027** **2029** 2031 2033 2035 2037 **2039** 2041

2043 2045 2047 2049 2051 **2053** 2055 2057 2059 2061 **2063** 2065 2067 **2069** 2071

2073 2075 2077 2079 **2081** **2083** 2085 **2087** **2089** 2091 2093 2095 2097 **2099** 2101

2103 2105 2107 2109 **2111** **2113** 2115 2117 2119 2121 2123 2125 2127 **2129** **2131**

2133 2135 **2137** 2139 **2141** **2143** 2145 2147 2149 2151 **2153** 2155 2157 2159 **2161**

2163 2165 2167 2169 2171 2173 2175 2177 **2179** 2181 2183 2185 2187 2189 2191

2193 2195 2197 2199 2201 **2203** 2205 **2207** 2209 2211 **2213** 2215 2217 2219 **2221**

2223 2225 2227 2229 2231 2233 2235 **2237** **2239** 2241 **2243** 2245 2247 2249 **2251**

2253 2255 2257 2259 2261 2263 2265 **2267** **2269** 2271 **2273** 2275 2277 2279 **2281**

2283 2285 **2287** 2289 2291 **2293** 2295 **2297** 2299 2301 2303 2305 2307 **2309** **2311**

2313 2315 2317 2319 2321 2323 2325 2327 2329 2331 **2333** 2335 2337 **2339** **2341**

2343 2345 **2347** 2349 **2351** 2353 2355 **2357** 2359 2361 2363 2365 2367 2369 **2371**

2373 2375 **2377** 2379 **2381** **2383** 2385 2387 **2389** 2391 **2393** 2395 2397 **2399** 2401

2403 2405 2407 2409 **2411** 2413 2415 **2417** 2419 2421 **2423** 2425 2427 2429 2431

2433 2435 **2437** 2439 **2441** 2443 2445 **2447** 2449 2451 2453 2455 2457 **2459** 2461

2463 2465 **2467** 2469 2471 **2473** 2475 **2477** 2479 2481 2483 2485 2487 2489 2491

2493 2495 2497 2499 2501 **2503** 2505 2507 2509 2511 2513 2515 2517 2519 **2521**

2523 2525 2527 2529 **2531** 2533 2535 2537 **2539** 2541 **2543** 2545 2547 **2549** **2551**

2553 2555 **2557** 2559 2561 2563 2565 2567 2569 2571 2573 2575 2577 **2579** 2581

2583 2585 2587 2589 **2591** **2593** 2595 2597 2599 2601 2603 2605 2607 **2609** 2611

Now, whilst, by the process described earlier, the distribution of the primes has been somewhat simplified, even a cursory inspection of the foregoing resembles the careless sprinkling of condiment over food. However it does show that, with the exception of the numbers **3, 5,** and **7,** it is not possible to have three consecutive prime numbers.

We now proceed to redress the seeming randomness of the distribution of the primes in the foregoing data. .

Cage numbers and the Magic Matrix

We have, somewhat jocularly described the matrix, Set1, that we are about to develop as, 'magic', but, it contains a plethora of interesting relationships between the numbers therein. Indeed, it is virtually certain that there are relationships that the author has not recognised. For example, the reader is invited to explore the horizontal and diagonal summation of the elements.

If the natural number line is examined it will be seen that there are groups of four numbers throughout which are bounded by numbers which are all divisible by three; we refer to these numbers as cage numbers. Hence, we have: **3,4,5,6; 6,7,8,9; 9,10,11,12; 12,13,14,15; 15,16,17,18**, the first five groups.

We now proceed to write these groups and their following versions in the manner shown:

3   4   **5**   6

9   8   **7**   6 --------------- [1]

9   10   **11**   12

15   14   **13**   12 --------------- [2]

15   16   **17**   18

21   20   **19**   18 --------------- [3]

21   22   **23**   24

27   26   25   24 --------------- [4]

27   28   **29**   30

33   32   **31**   30 --------------- [5]

33   34   35   36

39   38   **37**   36 --------------- [6]

39   40   **41**   42

45   44   **43**   42 --------------- [7]

45   46   **47**   48

51   50   49   48 ------------------ [8]

51   52   **53**   54 ------------------ [9]

Set1

7

It will be noted that, with the exception of **25, 35, 49** and **55**, all of the numbers in column 3 of this limited part of the matrix which are underlined and shown in bold are prime. If we continue to write the cage number groups in this fashion it will be seen that, with the exception of **2** and **3** we have embedded all of the prime numbers into column 3 of the matrix. The numbers to the right of the fourth column, shown in the square braces, we call the group numbers and each group contains two lines of the matrix.

When compared with the seemingly random distribution of the primes shown in the previous arrays we consider what has been done here to be a major simplification.

Further, if we add the group numbers in the form shown we get:

$1 + 2 = \underline{\mathbf{3}}$

$2 + 3 = \underline{\mathbf{5}}$

$3 + 4 = \underline{\mathbf{7}}$

$4 + 5 = 9$

$5 + 6 = \underline{\mathbf{11}}$

$6 + 7 = \underline{\mathbf{13}}$

$7 + 8 = 15$

$8 + 9 = \underline{\mathbf{17}},$

$9 + 10 = \underline{\mathbf{19}}$

$10 + 11 = 21$

$11 + 12 = \underline{\mathbf{23}}$

$12 + 13 = 25$

$13 + 14 = 27$

$14 + 15 = \underline{\mathbf{29}}$

$15 + 16 = \underline{\mathbf{31}}$

$16 + 17 = 33$

$17 + 18 = 35$ etc.   **Set2**

If we subtract the group numbers of column 1 in Set1 from the set of numbers on the right of the equality in Set2, many of the non-primes disappear. This is only one example that arises from use of the 'Magic matrix'.

8

Further, it is of passing interest to note that if we imagine the matrix to be laid out on a Euclidean plane, then, if we rotate the matrix in any sense through π radians about either of the first or last columns, then all of the primes in this instance are embedded in the second column of the resulting matrices.

We may determine the location of any number in the matrix; the following example illustrates the process:

Consider the number **23** in Set 1. Without inspecting the matrix we can determine that the nearest cage number must be **24**. If we divide this by **6** and multiply it by **2**, this will yield the group number with which the number **23** is associated, but this process always yields the last row of the matrix for this group, i.e., **8**; given the order in which the matrix is constructed it follows that the number **23** must lie in row **7**. Hence, in the usual notation of locating an element of a matrix, the coordinates of the number **23** are **(7, 3)**.

This reasoning may be applied to any number, for the matrix contains all of the natural numbers.

Strings and things

We define a primary string of numbers to begin and end with a number divisible by **6**. Hence, with reference to Set1, two primary strings selected at random are: **6 7 8 9 10 11 12** and **36 37 38 39 40 41 42**. If we examine these we find that the only numbers therein which can be prime are located in the second and penultimate positions on the string. Hence, since these strings were extracted at random we posit that this is a general feature of all such strings. In all cases the strings are fashioned from the last row of a group and the first row of the following group.

As an example of the use of the primary string, we choose the number **2407** from the second array shown on p6.

The nearest number to this which is divisible by **6** is **2406**, and, since this is less than the number that we have chosen, then the string with which this number is associated must be

2406 2407 2408 2409 2410 2411 2412. We need only examine the second and penultimate numbers in this series. Indeed, even by a cursory inspection of the numbers therein, it is obvious that this must be the case.

We now write out some of the strings in the sequence in which they arise in the Magic matrix:

6	**7**	**8**	**9**	**10**	**11**	**12**	$(k = 0)$	$(7 + 11)/6 = 3$
12	**13**	**14**	**15**	**16**	**17**	**18**	$(k = 1)$	$(13 + 17)/6 = 5$
18	**19**	**20**	**21**	**22**	**23**	**24**	$(k = 2)$	$(19 + 23)/6 = 7$
24	**25**	**26**	**27**	**28**	**29**	**30**	$(k = 3)$	$(25 + 29)/6 = 9$
30	**31**	**32**	**33**	**34**	**35**	**36**	$(k = 4)$	$(31 + 35)/6 = 11$
36	**37**	**38**	**39**	**40**	**41**	**42**	$(k = 5)$	$(37 + 41)/6 = 13$
42	**43**	**44**	**45**	**46**	**47**	**48**	$(k = 6)$	$(43 + 47)/6 = 15$

Set3

This selection is adequate for our purpose. We call **k** the string number. The reader will see that if we add the second and penultimate numbers in any string and divide by **6,** then we obtain one of the odd numbers, which we denote by **N.** Further, it is easy to see that we may determine any **N** from the simple formula; **N = 3 + 2k.**

The key to locating the prime numbers

If we examine the bounding numbers for any primary string it is seen that they are all members of the **6x** table. In addition, we conjecture that all of the prime numbers are embedded in the third column of the Magic matrix throughout the range of the counting numbers. Further, the disposition of the second and penultimate numbers of a string leads to confirmation of the prime number generating function, $p = 6q \pm 1, q = 1, 2, 3, etc.$, although, from inspection of Set3 we see that it is not infallible.

It then follows that if we wish to locate prime numbers then we may dispense with the strings and simply employ the **6x** table. Prime numbers can only exist on 'either side' of any member of this table---and nowhere else in the range of the counting numbers. For brevity we call these numbers, the associated numbers.

A simple illustration follows:

Let us take, at random, **6 x 391 = 2346.** The numbers immediately preceding and following this are **2345** and **2347.** The first of these cannot be prime for it ends in **5.** As shown in the preceding data, the second number is prime.

As a matter of interest, had we chosen **6 x 397 = 2382,** then we can see that the associated numbers, **2381** and **2383** are both prime. It is shown later in the work how we would determine the primality of these numbers. The location of associated numbers does not guarantee that these are prime, indeed, none of them may be prime, but this, in itself is valuable, for it contributes towards determining the gap between the primes.

Examining odd numbers for integer square roots and primality

As noted before there are many odd numbers that are obviously not prime, in that they have factors. However, some odd numbers have no obvious factors but do have an integer square root.

We now address the situation of determining whether any given odd number, N, has an integer square root. All of the numbers in the following discourse are, of course, integers.

Let a number N be given by: $N = c^2 = (a + b)^2$; hence $c^2 = (a + b)^2 = a^2 + 2ab + b^2$.

We now set, $b = a + 1$, and so, $c^2 = [2a(a + 1)] + [2a(a + 1) + 1]$.

From this we see that we may express the number of interest, i.e., N, as the sum of two consecutive numbers--- but this can only be done where the subject of the expression is an odd number, for it is not possible to express an even number as the sum of two consecutive numbers. If we take $N = P + Q$, then $P = 2a(a + 1)$, and hence, since we know N (which must be an odd number) we can write it in the form shown and we need only find two consecutive numbers, which when multiplied together and then by 2 will give P. This is illustrated as follows:

For brevity, we write $P = 2ab$. Consider the number, 81. This may be written as the sum:

$40 + 41 = 81$. Hence, $2ab = 40$, i.e. $ab = 20$, and, the only combination of consecutive numbers which can be multiplied together and yield 20 is, 4, 5. The sum of these two numbers is the square root of 81, i.e. $4 + 5 = 9$.

It may be inferred from this that if we form the products of 1,2; 2,3; 3,4; 4,5; 5,6; etc., then twice the product of each pair will give 2ab; if we then add 2ab + 1 the result will be the number, of which the sum of the digits of the product ab is the square root.

For example $2 \times 5 \times 6 = 60$, and, $60 + 61 = 121$, but the square root of 121 is $11 = 5 + 6$.

This procedure simplifies the determination of whether any given number, N, has a square root, for, if we write $N = P + Q$ where P and Q are consecutive numbers and P is an odd number, then N cannot have an integer square root (and hence a factor of this nature). If P is an even number and yields an odd number upon division by 2, then again, N cannot have an integer square root, for we cannot multiply two consecutive numbers and get an odd number. We may state succinctly that, if the number P is not given by twice the product of two consecutive numbers then the number N cannot have an integer square root.

Although it should go without saying, any odd number, N can be expressed as the sum of two consecutive numbers and so all the foregoing has universal applicability. It follows from this, that we can establish the following theorem:

If N is an odd number, and $(N - 1)/4 \neq a.b$, where a and b are consecutive numbers, smaller than N, then, $N \neq (a + b)^2$, and hence does not possess an integer square root.

12

A corollary of this is that we can generate all of the odd numbers which have a square root, for we may write: $N = 4ab + 1$.

We now set out two methods for determining the primality of a number.

It has already been shown that any odd number N can be expressed in the form $N = P + Q$, where P and Q are consecutive numbers; we may then write that $N = P + P + 1$. If we now divide throughout by an odd number, d, we get: $N/d = P/d + P/d + 1/d$. If we now complete the divisions on the RHS, there results: $N/d = A + n/d + A + n/d + 1/d$, where A is an unyet- specified integer and $n < d$. If we stipulate that n can take any value amongst the integers, then the sum of the fractions on the RHS must be unity and this will be accomplished for all, d, if we set $n = (d - 1)/2$. All of the foregoing is best illustrated by an example.

Consider the number 217. We may write this as $108 + 108 + 1$. If we divide 108 by 3 (i.e. $d = 3$) we get 108/3; but this should equal $A + 1/3$; hence, $A = (108 - 1)/3$, but 107/3 is not an integer and hence we may state that 3 is not a factor of 217. If we now set $d = 7$ then $n = 3$, and $A = (108 - 3)/7 = 105/7 = 15$ and so 7 is a factor of 217. If we sum the A's we get 30 and then add the unity which results from setting $n = 3$ and $d = 7$ in the expression for the sum of the fractions, i.e. $(2n + 1)/d$, we obtain 31, the other factor of 217. At this stage it could be argued that it would have been just as easy to divide N by 7, but in the process, in the absence of the equation derived earlier, i.e. $(N - 1)/4 \neq ab$, we would have lost the facility to test, initially if 217 had an integer square root (which, it does not).

We can write the above as a general process as follows:

Setting $n = (d - 1)/2$, the following table is generated:
Note that the ratio n/d tends toward 0.5*.

d	n	n/d & (decimal equivalent)	
3	1	1/3	(0.333333)
7	3	3/7	(0.428571)
11	5	5/11	(0.454545)
13	6	6/13	(0.461538)
17	8	8/17	(0.470588)
19	9	9/19	(0.473684)
23	11	11/23	(0.478260)
29	14	14/29	(0.482758)
31	15	15/31	(0.483871)
37	18	18/37	(0.486486)
41	20	20/41	(0.487805)
43	21	21/43	(0.488372)
47	23	23/47	(0.4893617)

If we divide $(N - 1)/2$ by any of the d in the above table and any resulting fraction corresponds to that associated with that d, then that d is a factor of N. It may be noted that even this modest table allows the investigation of the numbers in a range up to, roughly 2500.

* It is noted that the values in the last column of the table shown previously are tending towards **0.5** . Since it is expected that a calculator may sometimes be employed in the search for the prime numbers then this would display a whole number followed by the decimal part which may arise upon division of any number by a lower prime number. For large numbers this could present a problem in discriminating between the integer values of the numerator of the fractional part, **n** of the calculation. This may be obviated by the following procedure, which is also less complicated than that shown previously.

We choose the number $N = 217$ (which, obviously, has factors **7** and **31**, but this is of no matter here). Using the procedure developed before, if we divide **108** by **29**, for example we get **3.7241379**. It is not immediately clear what fraction the decimal part represents, but we can easily determine this if we subtract the whole part of the number and multiply the decimal part by **29**. This gives **20.999999**, which rounds to **21**. If, **29** was a factor of **217** then the number $14 = (29 - 1)/2$ would have appeared, hence we conclude that **29** is not a factor of **217**. Further, if we divide **108** by **7** we get $15.428571 = 15 + 0.428571 = A + 0.428571$, and, **0.428571 x 7 = 2.999997**, which we take as **3**, and which, in turn equals $(7 - 1)/2$. Hence we conclude that **7** is a factor of **217**. Also, another factor is given by $2A + 1$, **i.e. 2 x 15 + 1 = 31**.

Now, we have shown that $N = 3 + 2k$, and also that $N = 2P + 1$.

From this it follows that $P = 1 + k$.

We now examine the factor $1 + k$ in the search for prime numbers, in the process determining properties of k, by which we may eliminate particular numbers from primality. We intend to be quite pedantic in showing the status of the numbers that we investigate, even although, in some instances, that status is obvious. In addition we define a prime number as one which is only divisible by itself and unity. This is not trivial for as noted by Zagier [3] there are other definitions of what constitutes a prime number.

With reference to Set3:

$(k = 0)$, $(1 + k) = 1$. $P = (1 + k) = 1$, hence $N = 2P + 1 = 3$.

$1/3 = n/3$ and, $n = (d-1)/2$, hence 3 is a factor of N, and this is the only factor of N for there is no other d which will yield $n = 1$. Hence 3 is a prime number.

$(k = 1)$, $(1 + k) = 2$. $P = (1 + k) = 2$, hence $N = 2P + 1 = 5$.

$2/5 = n/5$ and, $n = (d-1)/2$, hence 5 is a factor of N, and this is the only factor of N for there is no other d which will yield $n = 2$. Hence 5 is a prime number.

$(k = 2)$, $(1 + k) = 3$. $P = (1 + k) = 3$, hence $N = 2P + 1 = 7$.

$3/7 = n/7$ and, $n = (d-1)/2$, hence 7 is a factor of N, and this is the only factor of N for there is no other d which will yield $n = 3$. Hence 7 is a prime number.

$(k = 3)$, $(1 + k) = 4$. $P = (1 + k) = 4$, hence $N = 2P + 1 = 9$.

$4/3 = 1 + 1/3$ and, $n = (d-1)/2$, hence 3 is a factor of N, but, 4/9 has a numerator which also accords with the criterion $n = (d - 1)/2$ and so $N = 9$ has two factors and hence 9 is not prime.

$(k = 4)$, $(1 + k) = 5$, $P = (1 + 4) = 5$, hence $N = 2P + 1 = 11$. Using the same procedure of trial division we see that the only fraction which qualifies for consideration is 5/11. As before it is seen that 11 is the only factor of 11, and so 11 is a prime number.

$(k = 5)$, $(1 + k) = 6$, $2P + 1 = 13$, and hence the same comments as that immediately above apply in this case, and so 13 is a prime number.

$(k = 6)$, $(1 + k) = 7$, $P = (1 + 6) = 7$, hence $N = 2P + 1 = 15$.

Now, $7/3 = 2 + 1/3$. The numerator of the fraction accords with $n = (d - 1)/2$ and hence 3 is a factor of 15. Also $7/5 = 1 + 2/5$; the numerator accords with $n = (d - 1)/2$ and so 5 is a factor of 15, hence 15 is not a prime number.

If we were to continue this procedure the following characteristics would emerge:

If k is divisible by 3 the associated N is not a prime number.

If k ends in 1 the associated N is not a prime number.

If $(1 + k)$ ends in 7 the associated N is not a prime number.

If $(1 + k)/2 = ab$, where a and b are smaller consecutive numbers, the associated N is not a prime number. We illustrate this by considering the number 49. Here, $2P + 1 = 49$ and so $P = 1 + k = 24$, and $(1 + k)/2 = 12$. But 12 is 4 x 3 and so 49 has a square root of $4 + 3 = 7$.

The reader is invited to continue using these procedures which will identify more prime numbers.

Further, we may carry out the foregoing analyses anywhere in the ranger of the counting numbers.

We conclude this section by referring back to Veisdal's statement in the Introduction.

This statement was preceded by a paragraph, the last line of which contained a string of prime numbers which terminated in **43**. He then asked the question 'what is the next prime number'.

In the light of the above we answer this as follows:

With reference to **Set3** it is seen that **43** is an associated number, and, the only possible place in the string which could contain a prime number is the penultimate place containing the number **47**. It must be borne in mind that the **k** associated with **47** is not the **k** of the set in which **47** appears, but rather the **k** determined from $(N - 1)/2 - 1$, i.e. $(47 - 1)/2 - 1 = 22$. Now, none of the constraints above apply to this this number, and so there is a strong possibility that it is prime. Hence we must use the version of trial division shown above:

$(1 + k) = 23$, $23/3 = 7\ 2/3$, $23/5 = 4\ 3/5$, $23/7 = 3\ 2/7$. None of the numerators satisfy the relationship $n = (d - 1)/2$, hence we conclude that **47** is the next prime number after **43**.

Discussion

Whilst not subscribing to Veisdal's excessive hyperbolae it is considered that the work presented here is a contribution to the investigation of the prime numbers and their distribution. The confining of the prime numbers to a single column of the Magic matrix, in conjunction with the concept of the primary string and the prime number generating function has revealed that prime numbers, instead of being distributed at random throughout the range of the counting numbers may only be found at well-designated locations---and nowhere else in that range. Further, we have established the veracity of a set of novel processes for determining primes, and since we make no distinction relating to types of prime numbers, our findings have universal applicability.

W M Fidler

May 2023.

References

[1] The Riemann Hypothesis explained

Jorgen Veisdal

Cantor's Paradise, Aug 21, 2016.

[2] On the determination of the primality of a number by use of an accelerated version of trial division.

W M Fidler

Grin Verlag, May 2021, ISBN 9783346493002.

[3] The first 50 million prime numbers.

Don Zagier

Inaugural Lecture

Bonn University, May 1975.

YOUR KNOWLEDGE HAS VALUE